BODY SYSTEMS: Human Cells

BY GARY RUSHWORTH

Table of Contents

Introduction	2
CHAPTER 1: Cell Structure	4
CHAPTER 2: Cell Function	12
CHAPTER 3: Cell Types	17
CHAPTER 4: Cell Organization	26
Conclusion	29
Glossary	31
Index	32

Introduction

Cells—do you know what they are? Or why they are important?

Do you ever wonder how you can do the things you do? Do you ever think about why you can see, smell, and taste; why you can run, jump, or ride a bike; or why you can think and learn? The answers have to do with the cell—the building block of all living things.

Humans, plants and other animals are all composed of cells. Cells perform all of the functions that keep organisms alive.

As you read this book, you will learn about the structure and function of cells. You will also learn about different types of cells and how they work together. Different types of cells work together to keep our bodies alive. Read on!

Chapter 1

Cell Structure

Cells are very small, yet very powerful. They are so small that you need a special tool to see them. This tool is called a microscope. Microscopes contain powerful magnifying lenses that help you see what cells look like by enlarging them.

There are about one trillion cells in the human body. One trillion looks like this: 1,000,000,000,000. These cells are classified into about 200 different types. Cells have different shapes, sizes, compositions, and functions.

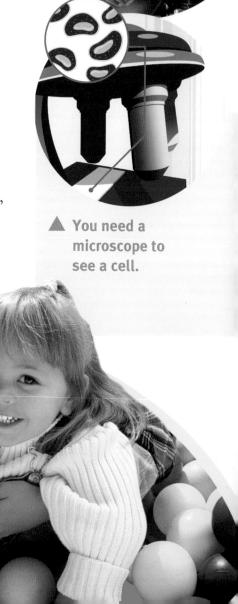

▲ You need a microscope to see a cell.

Each cell is about 60% (a little more than half) water. The remaining 40% is composed of structures, as well as chemicals, that a cell needs to function and sustain life.

The cell membrane keeps the contents of a cell intact. The cell membrane also acts as a barrier between the inside of the cell and the environment outside the cell.

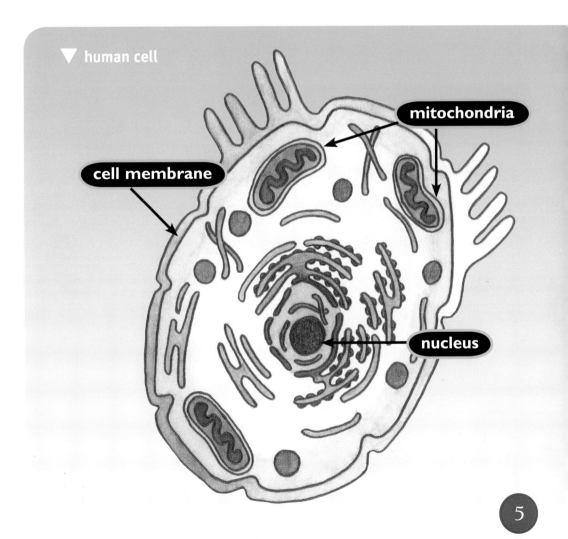

CHAPTER 1

This is an important role for the cell membrane because it protects the cell.

The cell membrane controls the movement of substances into and out of the cell. It allows nutrients to enter and wastes to exit. The cell membrane is able to do this because it has tiny holes in its surface called pores. The pores allow the cells to absorb water and nutrients and eliminate wastes.

It's a Fact

Plant cells have a cell wall. The cell wall does the same thing the cell membrane does—it holds things together. A cell wall, however, is stiff. The cell wall gives plants their shape. This allows the plants to grow upright and get the sunlight they need to make food.

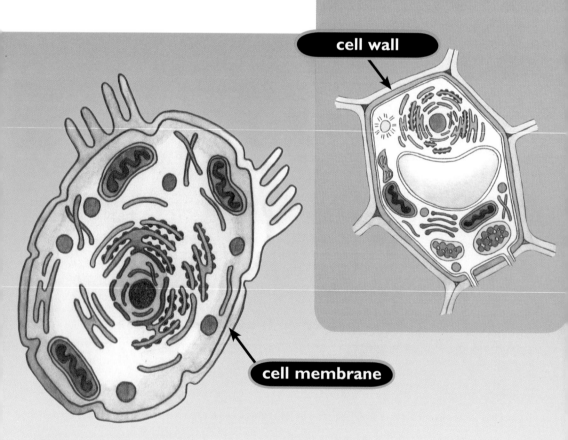

The cell membrane is sometimes described as a semipermeable membrane. Permeable means allowing substances to move through. Because the cell membrane is selective about what moves through, it is called semipermeable.

The **cytoplasm** (SY-tuh-pla-zum) is a jelly-like substance outside the nucleus and inside the cell membrane. The cytoplasm gives the cell its shape, anchors its parts, and controls the movement of those parts. Ribosomes and mitochondria are structures found in the cytoplasm.

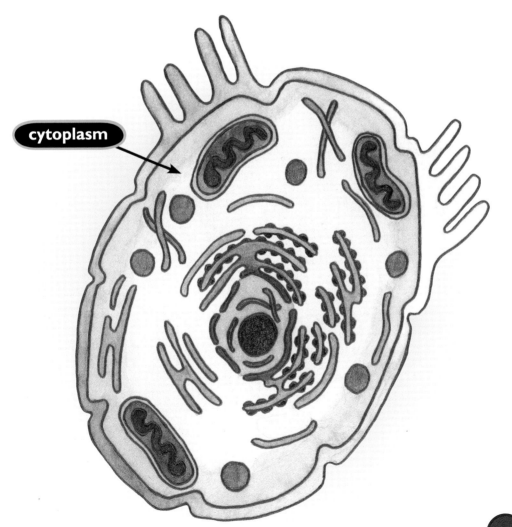

CHAPTER 1

The **nucleus** (NOO-klee-us) can be described as the brain, or control center, of the cell. The nucleus controls all of the functions of the cell. It decides what kind of cell to make and even determines when more cells are needed.

The nucleus is controlled by **genes**. Genes are substances that make up part of the body's chemical code. This chemical code contains all the instructions and information the nucleus uses in deciding what kind of cell to make. The code determines how the cell develops, grows, and maintains and repairs itself.

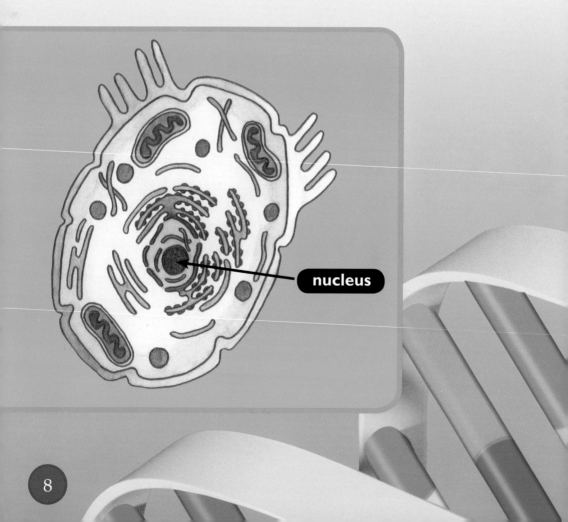
nucleus

CELL STRUCTURE

Genes are segments, or lengths, of DNA, or deoxyribonucleic acid. DNA is a complex chemical substance shaped like a double helix, or twisted ladder. The order of chemicals in a DNA segment determines the information the gene carries.

It's a Fact

More than 99% of your DNA is the same as every other person. We are really more alike than different. The differences make us the individuals we are.

▲ a DNA strand

CHAPTER 1

The DNA strands making up the genes are part of a larger unit called a **chromosome** (KROH-muh-some). Chromosomes are arranged in pairs inside the nucleus. Each chromosome can contain more than one thousand genes. The information chromosomes contain tells the cell what to do. The information gets passed along as new cells form. It also passes from generation to generation. You get this code from your parents, and you pass it on to your children.

▼ chromosomes

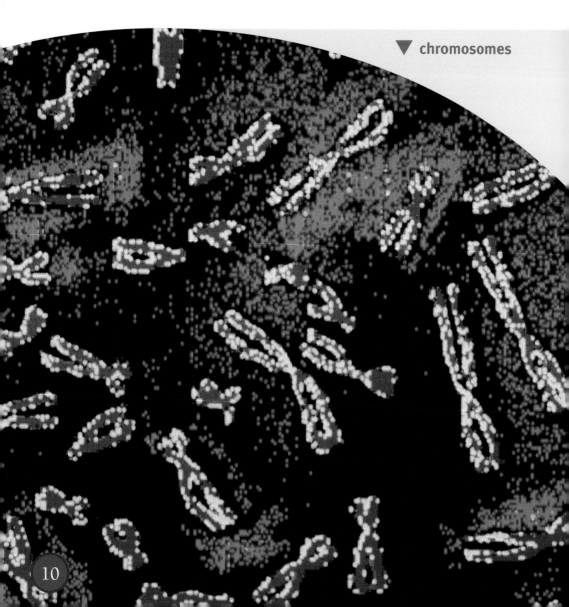

CELL STRUCTURE

There are other structures inside the cell. They are called **organelles** (or-guh-NELZ). Organelles are located in the cytoplasm and do specific tasks to make the cells work.

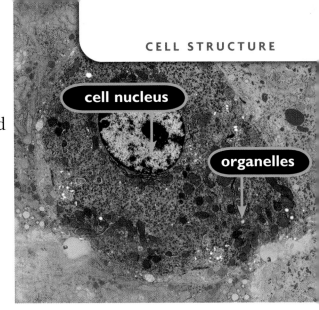

Major Cell Structures and Organelles

Cell Part	Description
Cell Membrane	A layer surrounding the cell that holds the cell's contents together and controls the movement of materials into and out of the cell
Nucleus	Control center of the cell that contains the cell's DNA
Cytoplasm	Watery fluid inside the cell membrane and outside the nucleus that suspends the organelles
Mitochondria	The "powerhouses" of the cell that break down food to make energy for cell function
Vacuoles	Store water and other nutrients to be used later by the cell
Endoplasmic Reticulum	Nutrient transportation system within the cell

Chapter 2
Cell Function

Cells do many things to make sure our bodies function well and stay healthy. Cells need fuel to provide energy to do the work they do. They also need lots of water. Cells could not survive without **oxygen**. They get these substances from what we eat, drink, and the air we breathe.

The food is broken down into a sugar called **glucose** (GLOO-kose). Water and oxygen are absorbed into the cell and combine with the glucose to make energy. This energy production occurs in the mitochondria (my-tuh-KAHN-dree-uh).

Cell Energy

The mitochondria are organelles that turn the glucose into energy. It is for this reason that they are called the "powerhouses" of the cell. The mitochondria absorb the fuel they need from the food we eat.

Before the mitochondria can absorb the food we eat, the food must be broken down. This is done by the **lysosomes** (LY-suh-somez). Lysosomes contain **enzymes** (EN-zimez). Enzymes are chemicals that break down the food we eat into a form the mitochondria can use. The mitochondria use enzymes to break down waste products, too. The waste goes into the cytoplasm. This waste is then eliminated from the cell.

◀ Mitochondria are the "powerhouses" of the cell.

Stomach and Associated Cell Tissues

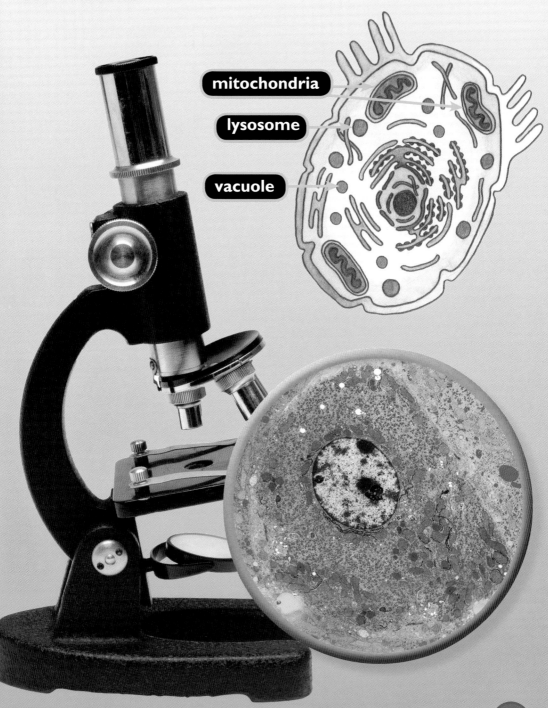

Glucose, water, and oxygen enter the cell by a process called diffusion. Unwanted waste products are eliminated in the same way.

Diffusion occurs across the cell membrane. Material flows from where there are more particles to where there are fewer particles. Diffusion stops when the concentrations of particles are equal. Because there are more particles of glucose, water, and oxygen outside the cell, they flow into the cell. This brings the mitochondria the fuel it needs to make energy for the cell's processes.

Reread

Reread page 14. What is diffusion and when does it occur?

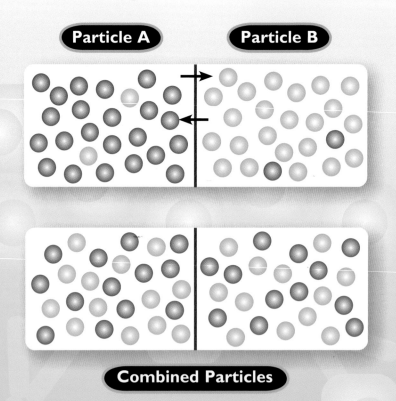

CELL FUNCTION

Once inside the cells, nutrients and water are spread throughout the cytoplasm. Some of these nutrients and water are used by the cell right away to make energy and perform the cell's functions. Some are not.

If the nutrients and water are not used right away, they are stored for later use in the cytoplasm and in vacuoles (VA-kyuh-olez)—bubble-like structures found in the cytoplasm that act as a storage and transport container.

The processes by which a cell makes energy and eliminates wastes are parts of **metabolism** (muh-TA-buh-lih-zum). Metabolism is the sum of all the chemical activities a cell performs.

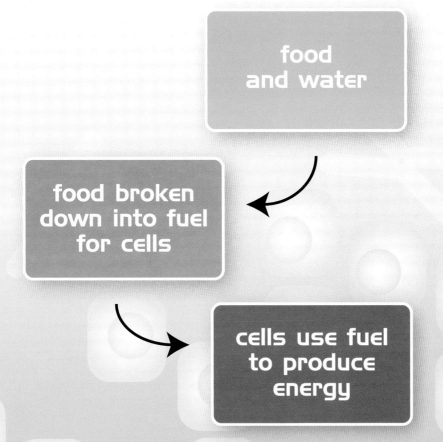

CHAPTER 2

Cells don't last forever. When cells are used up, they break apart, dissolve, and are eliminated as waste. The body must have the ability to make new cells when the old ones die. There is a constant need for new cells to replace the ones that die. Cells divide to make new cells. The process of cell division resulting in the formation of two identical cells is called **mitosis** (my-TOH-sis).

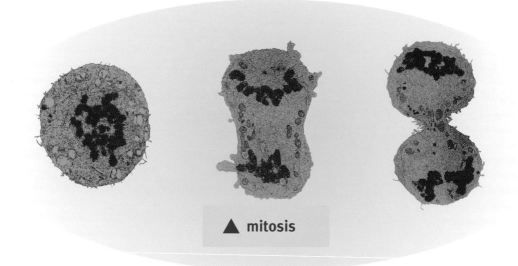

▲ mitosis

They Made a Difference

In 1839, German biologist Theodor Schwann said that all living things were made of cells, and that cells relied on each other to work together for the good of the whole living thing. Later that year, a fellow biologist, Rudolf Virchow, said that all cells come from other cells.

Chapter 3

Cell Types

We have learned what cells are made of. We have read about how cells get the energy they need to keep working and some of the jobs cells do. We have seen cells divide into two identical cells. We know cells can have similar functions or widely different ones. We know that there are about 200 different kinds of cells.

They Made a Difference

In 1655, the English scientist Robert Hooke discovered cells while examining a dried section of a cork tree with a microscope. He observed small chambers and named them cells. This discovery is where it all began.

Careers in Science

Biologists are people who study the structure and function of living things. There are plant biologists and animal biologists. Biologists who study just cells are called cellular biologists.

CHAPTER 3

Stem Cells

There are many kinds of cells, but they all start the same way. All cells grow from what is called a stem cell. A stem cell is a specialized cell that has the ability to develop into different kinds of cells in the body. When the body needs new cells, the stem cells start the process of making more of the kinds of cells that are needed.

Stem cells watch the cells around them as they grow in size and number. If there is not enough of one cell type, the stem cells make more of that type and less of another. When the number of each cell type is correct, the stem cells simply make more stem cells. Stem cells are also the repair shop for cells. They make new cells of all kinds and repair damaged cells, as well.

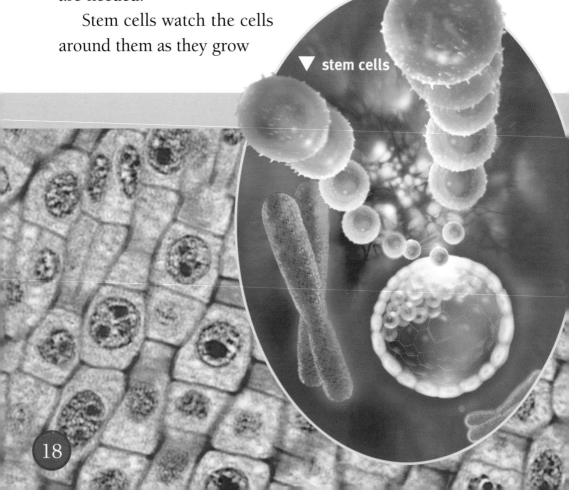

▼ stem cells

Skin Cells

Skin cells make up the covering of our bodies. Our skin is composed of billions of skin cells and is actually considered an organ—the largest organ of the human body. Skin cells protect the body the same way the cell membrane protects the cell. Skin cells regulate body temperature and body water. Skin cells are a barrier to dirt and disease.

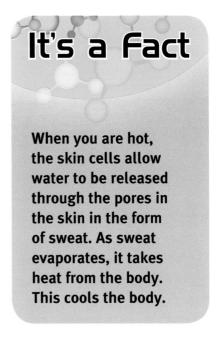

It's a Fact

When you are hot, the skin cells allow water to be released through the pores in the skin in the form of sweat. As sweat evaporates, it takes heat from the body. This cools the body.

▼ skin cells

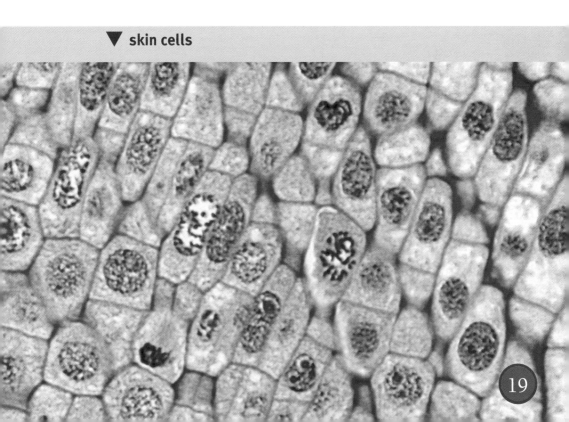

CHAPTER 3

Blood Cells

Blood is the fluid that flows throughout our bodies. Blood is made up of different kinds of cells that float in a fluid called **plasma**. The two most important blood cells are red blood cells and white blood cells.

Red Blood Cells

Red blood cells are the body's grocery stores. They carry food, nutrients, and oxygen to other cells. They take away waste materials to be eliminated from the body. Oxygen is one of the fuels body cells need. Oxygen comes from the air we breathe. Red blood cells deliver oxygen to all the body cells.

It's a Fact

Red blood cells are the only cells that do not have a nucleus.

▲ red blood cells

20

CELL TYPES

White Blood Cells

White blood cells are the body's clean-up crew. They protect the body from disease and help keep the blood clean. They also help to heal minor cuts and scrapes and prevent infection from bacteria and viruses.

When bacteria or viruses attack the cells, the white blood cells attach to them. Then they wrap around the bacteria or viruses. This keeps the bacteria or viruses from getting into the cells.

▲ white blood cells

21

Bone Cells

Bone cells are the cells that make your bones. Bones form the body's skeleton, or support system. Together with the muscles, bones allow us to stand, sit, and move about. They also protect the body's internal organs, such as the heart and brain, from damage.

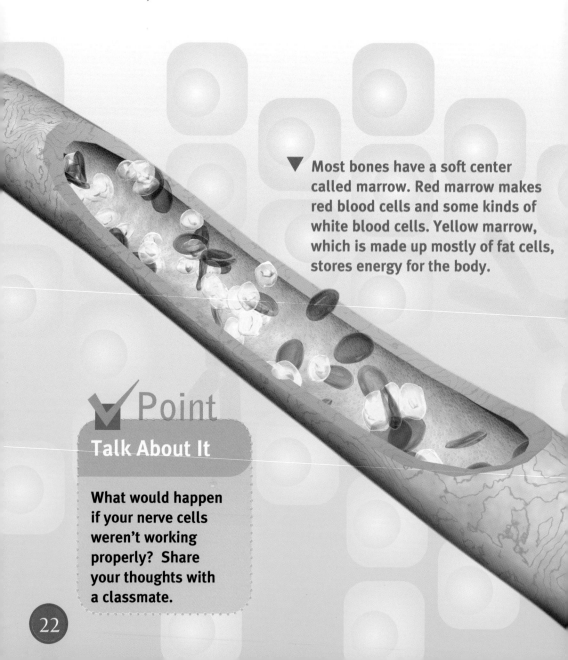

▼ Most bones have a soft center called marrow. Red marrow makes red blood cells and some kinds of white blood cells. Yellow marrow, which is made up mostly of fat cells, stores energy for the body.

✓Point
Talk About It

What would happen if your nerve cells weren't working properly? Share your thoughts with a classmate.

Muscle Cells

There are three basic types of muscle cells: striated muscle cells, smooth muscle cells, and cardiac muscle cells. Striated muscle cells are part of the large muscles in the body. These muscles help you walk, climb, run, push, and pull. They are called voluntary muscles because they are under your control as you walk, run, or throw a ball.

Smooth muscle cells are part of your internal organs, such as your stomach, intestines, and lungs. Smooth muscle is called involuntary muscle. It functions without your control. Cardiac muscle cells are found only in the heart. Cardiac muscle is involuntary.

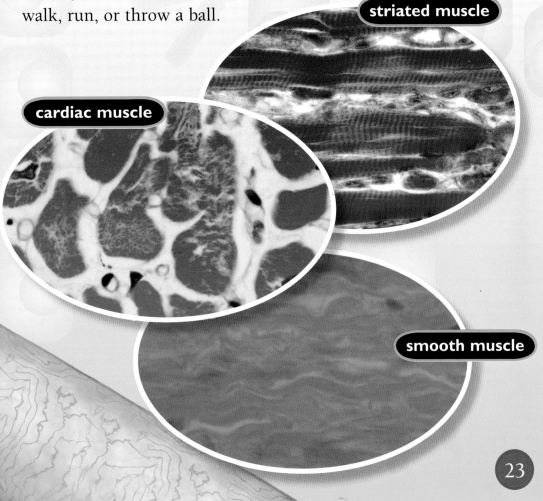

Nerve Cell (Neuron)

Nerve cells, called neurons, make up the brain, spinal cord, and nerves of the body. They send and receive "messages" to and from all parts of the body. The human brain has about 100 billion neurons. They tell the body what to do.

Nerve cells receive information from all the cells of the body. They can tell how much water is in the cells or how fast or slow they are working. They know when we need to eat and how hot or cold we are. They know when we are hurt. They also know when we are tired. All the information collected by the nerve cells goes to the brain. The brain tells the nerves to tell the cells what to do.

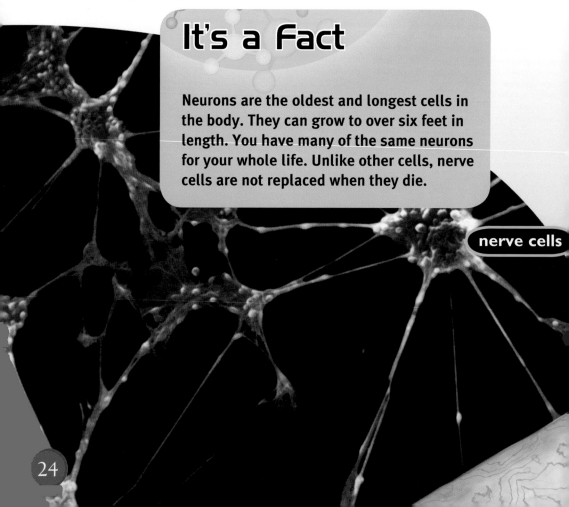

It's a Fact

Neurons are the oldest and longest cells in the body. They can grow to over six feet in length. You have many of the same neurons for your whole life. Unlike other cells, nerve cells are not replaced when they die.

nerve cells

These are just a few of the 200 different types of cells in the body. Together, they control who we are and what we do. They perform specific functions but also work together so that the body can be healthy and stay alive. Considering that the human body has more than one trillion cells, this is quite an accomplishment!

Cell Types

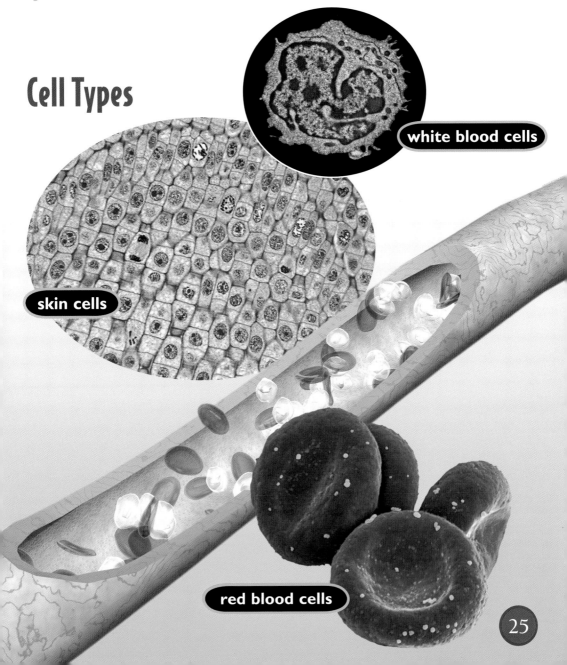

white blood cells

skin cells

red blood cells

Chapter 4

Cell Organization

The human body has about 200 different kinds of cells. These cells must work together to keep you alive. Each kind of cell has a specific structure and function.

A cell is always functioning. But one cell cannot perform all of the jobs needed to keep a complex organism alive.

Similar cells group themselves into a unit called a tissue. Tissues are organized groups of similar cells that work together to do a specific task. Muscle cells combine to form muscle tissue just as nerve cells combine to form nerve tissue.

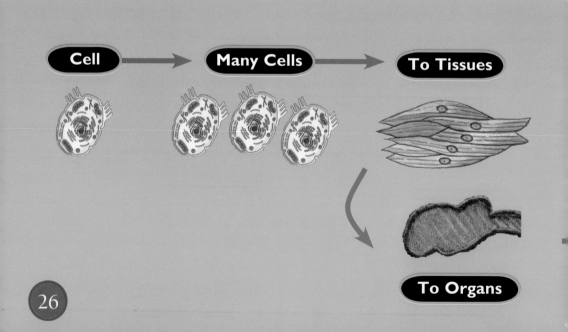

But tissues by themselves are still not enough to handle all the things our bodies need to do.

Sometimes different tissues work together to perform a specific function. Such a group of tissues is called an organ.

Examples of organs include the heart, brain, stomach, lungs, and skin. The skin is an organ because it is made up of different groups of tissues.

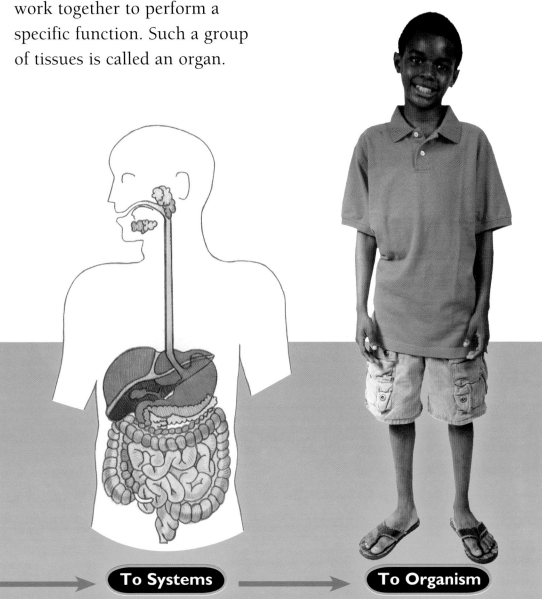

To Systems → To Organism

CHAPTER 4

Organs also work together in groups to perform certain jobs. These groups are called systems. Each system has a specific job to do. For example, the brain, spinal cord, and nerves form the nervous system. The nervous system carries information between the brain and other parts of the body. Other systems include the respiratory system and the digestive system. A system can function only as well as the organs that make it up. If one organ fails, the whole system is affected. A system that is not working properly can affect other systems. Finally, all the systems work together to make an organism—an organism such as you!

▲ The way your body works begins with your cells.

Conclusion

The human body is an amazing machine. We don't know everything about how it works yet, but we do know that it is made up of building blocks called cells. The human body is made up of more than a trillion cells. Each type of cell has a different function, or job.

Cells are incredible. What makes them so incredible is that they all work together and on their own to keep your body healthy and alive.

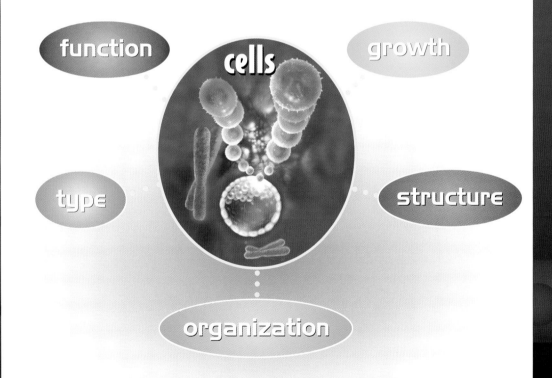

CONCLUSION

You have a responsibility to your body to keep it in good working order. As you know, cells need water, oxygen, and glucose to function. That means you need to take care of your body by eating a balanced diet, exercising, and getting enough sleep.

▲ Cells are incredible. Take care of them.

Glossary

cell	(SEL) the basic unit of all living things (page 2)
chromosome	(KROH-muh-some) a structure in the nucleus containing DNA (page 10)
cytoplasm	(SY-tuh-pla-zum) watery jelly-like substance that contains the organelles outside the nucleus (page 7)
enzyme	(EN-zime) a chemical that breaks down foods for the cells (page 12)
gene	(JEEN) a segment of DNA located on a chromosome that carries information passed from parent to offspring (page 8)
glucose	(GLOO-kose) a form of sugar the cells use to make energy (page 12)
lysosome	(LY-suh-some) organelle that contains enzymes (page 12)
metabolism	(muh-TA-buh-lih-zum) the sum of all the chemical activities a cell performs (page 15)
mitosis	(my-TOH-sis) cell division into two identical cells (page 16)
nucleus	(NOO-klee-us) the cell's control center (page 8)
organelle	(or-guh-NEL) part of a cell that performs a specific task (page 11)
oxygen	(AHK-sih-jen) chemical necessary for sustaining life (page 12)
plasma	(PLAZ-muh) the watery, liquid portion of blood (page 20)

Index

bacteria, 21

blood, 20–21

cell, 2–5, 10–12, 14, 16–22, 24, 26, 29

cell membrane, 6–7, 19

cell wall, 6

chromosome, 10

cytoplasm, 7, 15

enzymes, 12

gene, 8–9

glucose, 12, 14

lysosomes, 12

metabolism, 15

microscope, 4, 17

mitochondria, 12, 14

mitosis, 16

nucleus, 8

organ, 19, 27–28

organelles, 11–12

oxygen, 12, 14, 20

plasma, 20

red blood cells, 20

skin cells, 19

smooth muscle cell, 23

stem cells, 18

striated muscle cells, 23

system, 22, 28

tissue, 26–27

white blood cells, 21